MW01488546

Thunderstorms and Lightning

Dean Galiano

the rosen publishing group's

rosen central

new york

Published in 2000 by The Rosen Publishing Group, Inc.
29 East 21st Street, New York, NY 10010

Copyright © 2000 by The Rosen Publishing Group, Inc.

All rights reserved. No part of this book may be reproduced in any form without permission in writing from the publisher, except by a reviewer.

First Edition

Galiano, Dean.
        Thunderstorms and lightning / Dean Galiano.
            p.   cm. -- (Weather watchers' library)
        Includes bibliographical references and index.
        Summary: Explains the formation and activity of various kinds of thunderstorms as well as the electrical and physical properties of lightning. Includes a chapter on safety during storms.
        ISBN 0-8239-3093-9
        1. Thunderstorms Juvenile literature. 2. Lightning Juvenile literature. [1. Thunderstorms. 2. Lightning. 3. Weather.]
I. Title. II. Series: Galiano, Dean. Weather watchers' library.
QC968.2.G35 1999
551.55'4--dc21   99-33160
                                                       CIP

Manufactured in the United States of America

CONTENTS

Introduction

Weather is a topic you just can't ignore. It affects you every day. What you wear, how you travel, when you leave for school: Each is influenced by the weather, and may be determined by it. Six inches of snow on the ground means you might not be going to school, but you will be shoveling it off the driveway. A warm, sunny day requires light, comfortable clothing. And when it rains, you are forced to notice the weather. Sometimes when it's raining you may have to stay inside all day. Otherwise, rain is a good thing. It waters the trees and plants, helps our crops to grow, and supports all human life.

All of the weather that we experience takes place in something called the atmosphere. The atmosphere is a thin layer of air that surrounds Earth. When you look up into the huge sky, with its great masses of floating clouds, it is hard to think of the atmosphere as being small—but it really is. Weather actually happens in a very limited space. If Earth were shrunk to the size of a basketball, our atmosphere would be thinner than a piece of notebook paper!

Sometimes the weather is nice, sometimes it is dreary, and sometimes it can turn downright nasty. Thunderstorms are the most common form of violent weather. Thousands of them take place every day, all over the planet. Thunderstorms are big, powerful, and carry with them huge amounts of water and energy. Farmers depend on this water to provide moisture for their crops. The energy that thunderstorms produce, however, is capable of destroying almost anything in its path. They can produce violent winds that are able to blow over telephone poles, trees, and even homes. They can release huge amounts of rain in just a few minutes, causing dangerous flash floods, where streams and rivers overflow their banks. And they can drop hailstones as large as softballs, which speed to Earth at over 100 miles per hour. Thunderstorms also produce lightning, which is a major cause of forest fires and a danger to human life.

This is a cumulus cloud forming in the atmosphere.

1 Thunderstorm Formation

Thunderstorms are one of the most spectacular events experienced on Earth. Most of us are familiar with thunderstorm weather: The sky blackens, cool gusts of wind begin to blow, and suddenly, the air is filled with torrential rain and lightning flashes. Thunderstorms are indeed spectacular, yet they are the most common storms on Earth. It is estimated that 44,000 thunderstorms occur every day. In fact, at this moment nearly 2,000 thunderstorms are in progress somewhere on the planet.

Thunderstorms are the product of cumulonimbus clouds. These clouds look like huge, lumpy, snow-covered mountains floating in the air. Commonly called thunderheads, these clouds can cause great destruction as they pass overhead. When thunderheads release their energy, it is an awesome experience. Powerful strokes of lightning flash across the sky, followed by deafening

Thirsty crops await rain brought by a thunderstorm.

thunderclaps. Intense gusts of wind can uproot trees and even blow the roofs off of houses. Large hail may kill farm animals and damage cars and buildings.

There is no doubt that thunderstorms are capable of great destruction. Yet they are also vital to life in the United States. Without the moisture of the rain, many farms in the Midwest would not be able to produce crops. Also, much of the water that we rely on to drink comes from the rain that thunderstorms produce.

How Thunderstorms Form

Thunderstorms are born from cumulus clouds. These are the puffy white clouds that float gently across the summer sky. Cumulus clouds are formed when a mass of moist air comes into contact with a section of the ground that has been heated by the sun. Contact with the ground warms the moist air. The air then becomes lighter and floats upward into the atmosphere. As this warm, moist air rises into the atmosphere, it moves into areas where temperatures are lower. Eventually, the rising air cools to the dew point. The dew point is the temperature at which water vapor (water in the form of a gas) turns into a liquid through a process called condensation. The condensed water vapor forms water droplets, which are about a million times smaller than a single raindrop. These droplets are held in the air by the rising air mass. The rising air then becomes visible as a cloud.

The formation of cumulus clouds depends upon the presence of a large amount of moisture in the air. It also depends upon a temperature difference between the air near the ground and the air at higher levels in the atmosphere. When the temperature difference is small, and the moisture level is low, only small cumulus clouds form. Usually these clouds are

Cumulus clouds form as warm, moist air rises to meet cooler air temperatures.

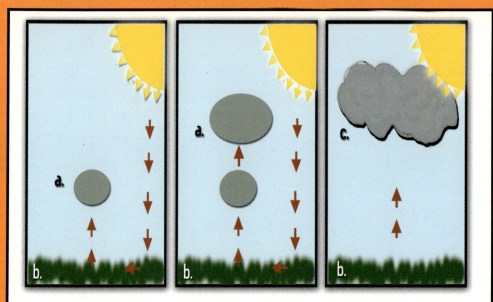

A moist air mass (a) reacts with the ground (b), which has been heated by the Sun, pushing the air mass (red arrows) up into the atmosphere, where it is cooled to its dew point to create clouds (c).

only a few thousand feet tall, well below the height at which jet airplanes fly. These clouds do not grow to greater heights because the two conditions they need for growth do not exist in the atmosphere above them. These conditions are:

1) high moisture levels; and
2) large temperature differences between the rising air and the surrounding air.

However, when these two conditions do exist at greater heights, these small, puffy clouds can grow into huge cumulonimbus clouds (thunderheads).

Heat and Moisture

For a mass of air to rise high enough into the atmosphere to cool to the dew point and form clouds, it must become lighter than the air surrounding it. There are two ways that air can become lighter:

1) Water vapor can be added to the air.
2) The air can be heated.

It may seem hard to believe that adding water to air can make it lighter, but it is true. This is because water vapor actually weighs less than dry air. Thus, when water vapor enters a parcel of dry air, it makes the air lighter. Air can also be heated to make it lighter. Think of a hot-air balloon. The reason the balloon floats is that the air inside of it is warmer, and therefore lighter, than the air outside it.

Expanding Air

As warm, moist air rises, it also expands (spreads out). This happens because as air rises into the atmosphere, it meets with lower air pressures. Air pressure is the weight of air in the atmosphere pressing down on any one spot.

A rising parcel of air expands under its own weight, just as a piece of clay would expand if you pushed down on it with your hand. When warm, moist air expands, it also cools. The air will continue to rise until it is the same temperature as the air around it. When the difference in temperature between the rising air and the surrounding air is small, the rising air will only form small cumulus clouds. If the difference in temperatures is large, though, the cloud will continue to grow. When the temperature difference is great, the air is called unstable.

Thunderheads, or cumulonimbus clouds, form when the air is unstable. Under such conditions, the temperature difference between the warm, moist air and the surrounding air becomes greater with increasing height. Therefore, the rising air mass inside the cloud rises faster and faster into the atmosphere. Think of the inside of the cloud as a fireplace chimney—as warm air moves up into the cloud, fresh air is drawn in from below. If there is a great deal of moisture and a great temperature difference between the rising air and the surrounding air, the cloud will continue to grow.

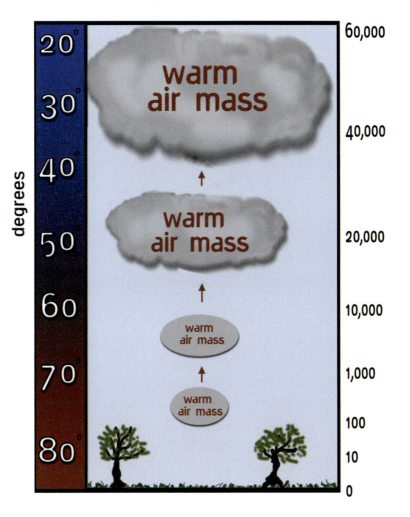

The higher an air mass rises, the greater becomes the difference in air temperature. This difference fuels rapid cloud growth.

Thunderstorm Development

Thunderstorms usually develop in one of two ways:

1) Warm, moist air is strongly heated by contact with a section of Earth's surface.
2) Masses of warm, moist air are pushed upward by masses of cooler air.

Thunderstorms that are produced by strongly heating Earth's surface usually happen on a summer afternoon. During the summer, the ground is warmed by the sun all day and is at its highest temperature. When a mass of moist air comes into contact with the heated surface, the air becomes warmer and begins to rise. If the temperature of the rising air is much warmer than the temperature of the surrounding air, cumulonimbus clouds will develop. These clouds may then develop into thunderstorms. Thunderstorms formed in this way are called air-mass thunderstorms.

Thunderstorms that develop along a front are called frontal thunderstorms. A front is an area where a mass of cold air meets a mass of warm air. When a cool, dry air mass meets a warm, moist air mass, the warm, moist air mass is pushed upward because it is lighter. As it rises into the atmosphere,

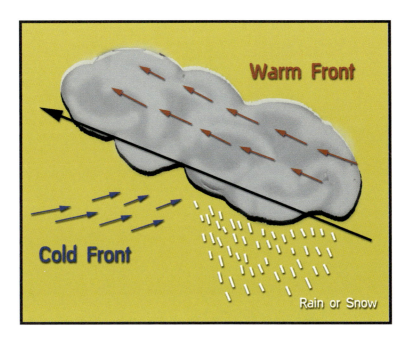

Frontal clouds form where a mass of warm, moist air meets and reacts with a mass of cooler air.

the water vapor in the air cools to the dew point. At this point, clouds form because the water vapor condenses. Frontal thunderstorms can happen at any time during the year, although they happen most often in the spring and summer.

There are three stages in the life of a thunderstorm: the cumulus stage, the mature stage, and the dissipating stage.

The Cumulus Stage

Thunderstorms develop from cumulus clouds. The cumulus stage lasts for about ten to fifteen minutes, during which cloud growth is rapid. The clouds grow tall as updrafts—strong winds that blow up from the ground—flow into the cloud to make it stronger as heat is added to the cloud. The added heat comes from water vapor: As water vapor condenses, it gives off heat. The more water vapor in the air, the greater the amount of heat given off. With help from the added heat, the cloud continues to grow rapidly. In just a few minutes, a small cumulus cloud—about a mile wide and 5,000 feet tall—can grow into a huge cumulonimbus cloud.

Cumulonimbus clouds are several miles wide and can grow as tall as seven miles high. During its growth, the cloud expands and rises into colder temperatures. Eventually it reaches into the freezing temperatures range. Above the freezing level, small particles of ice and snow

High levels of moisture collected in clouds can be seen by their darkness.

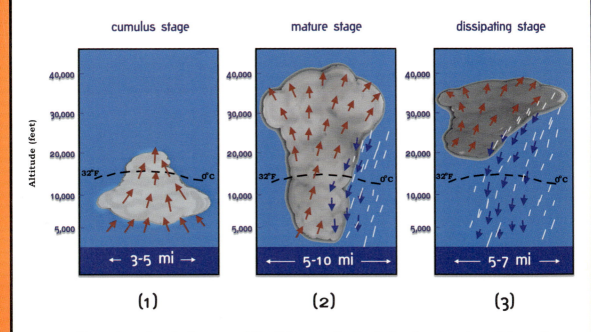

cumulus stage   mature stage   dissipating stage

(1)   (2)   (3)

Cumulus clouds form rapidly (1) as updrafts bring heat and moisture from the ground; a mature thunderstorm grows to its tallest level and flattens out (2), and the moisture collected becomes too heavy to hold, allowing its precipitation to fall to the ground; when the downdraft cuts off the supply of moist air flowing up into the cloud (3), the cloud begins to evaporate.

crystals combine to form larger particles. When these particles join, they release heat. This heat keeps the updrafts strong at higher and higher levels. As heat is released into the cloud, the air becomes lighter, and the air in the cloud's updraft rises more rapidly.

While the cloud is growing taller, rain, snow, and hail are held up in the cloud by the updrafts. These particles combine with each other and grow larger.

17

Soon the particles grow too large to be held up by the updrafts. They begin to fall, and precipitation begins. Precipitation is any water that falls from a cloud, such as rain, snow, or hail. As the precipitation falls, it drags down with it air from the lower part of the cloud. This downward pulling of air is called a downdraft, which is a strong wind blowing down against the ground from the atmosphere. Now the thunderstorm has reached the mature stage.

The Mature Stage

During the mature stage, thunderstorms grow to their tallest levels. They often reach as high into the atmosphere as 60,000 feet. This layer of air is called the stratosphere. Strong horizontal winds in the stratosphere flatten out the top of the cloud. This is why thunderheads have a flat, anvil-shaped top.

As precipitation begins, the way the cloud works becomes more complicated. As particles of rain and ice

A pedestrian wisely guards against the rain.

Heavy rains often bring floods.

fall through the cloud, many of them evaporate—
they turn back into water vapor. Evaporation cools
down the air. When air cools, it becomes heavier
and begins to sink. This adds to the force of a
cloud's downdraft.

Now the cloud has a strong updraft in one part
of it and a strong downdraft in another part. The
updraft is usually twice as strong as the downdraft.
The downdraft travels down to Earth's surface and
spreads out ahead of the storm. Have you ever

noticed that, just before a thunderstorm hits, there is a cool gust of air that scatters leaves about? This cool gust is the downdraft of the approaching storm. Raindrops follow soon after and fall more intensely as the storm moves overhead. The heaviest rain falls directly under the center of the storm. Heavy rains usually last for about ten to twenty minutes.

Although thundershowers usually last for less than thirty minutes, they deposit a large amount of rain. A thundershower can drop as much rain in fifteen minutes as a gentle rain will in a twenty-four-hour period.

At times during a thunderstorm, the updraft can suddenly decrease in force. When this happens, many of the raindrops that were being held up in the cloud fall at once. This is called a cloudburst. Thick sheets of rain pour downward. An inch or more of rain can fall in just minutes, causing flash floods. Such flash floods can wash away homes, bridges, and even roads. More than one hundred people drown in flash floods every year.

Hailstones

Hail is a type of precipitation. Hailstones are balls of ice that range from pea-sized pebbles to softball-sized chunks. Hail falls from intense thunderstorms that have strong updrafts and reach great vertical heights. A hailstone is formed when an ice pellet in a cumulonimbus cloud grows larger as it hits and combines with supercooled— below freezing temperature but still liquid—water droplets. The stronger the updraft in the cloud, the more time the hailstone has to grow. Eventually, the hailstone will grow too large and become too heavy to be supported by the updrafts. It then falls to the ground. Large hailstones can damage crops, dent cars, and even kill farm animals.

The Dissipating Stage

As a thunderstorm progresses, the downdraft spreading out along the surface of the Earth cuts off the supply of air flowing into the updraft. As the updraft is cut off, so is the supply of warm, moist air that builds and nourishes the huge clouds. The storm then enters the dissipating stage. At this point the cloud begins to evaporate. Soon the rain, lightning, and thunder stop. When the clouds dissipate, the sky clears and a new weather cycle begins.

2 Types of Thunderstorms

There is more than one type of thunderstorm. Thunderstorms are grouped by size and strength. Some thunderstorms are weak in power, lasting only a few minutes, and producing only a small amount of rain. Other thunderstorms can be powerful and dangerous, with flashing lightning, loud thunder, and huge amounts of rain coming down for hours. These different types of thunderstorms have been given names to identify their strength and violence.

Single-Cell Thunderstorms

Single-cell thunderstorms are small, usually weak thunderstorms that are often isolated (alone). They seem to grow from any warm, humid air mass in the area. But this is not really so. Usually, they show up on the front edge of a larger mass of thunderstorms that will soon follow. This is because they are in the same air mass as the larger weather

Thunderstorms stretch across the sky, threatening a residential neighborhood.

system behind them. Single-cell thunderstorms can also appear at the back of a larger thunderstorm cell that has ended its cycle. Such small thunderstorms have a life span of only 30 minutes. However, single-cell thunderstorms are not the most common type of thunderstorm.

Multicellular Thunderstorms

Most thunderstorms are multicellular, having many single cells grouped together. These are more powerful thunderstorms than single-cell storms. Such large thunderstorms cause loud thunderclaps and lightning flashes down to the ground. Heavier rains come from multicellular thunderstorms, and they can last many hours. This is because all the many cells in this group are usually at different stages of development. As one cell dissipates, another cell may just be maturing. Often this type of thunderstorm may be threatening an area miles ahead of its path, but by the time it reaches that area it has already grown very weak or dissipated. However, it is often hard to tell when a multicellular thunderstorm will grow weak and dissipate. It's possible for such storms to change back and forth throughout their life cycle, from weak thunderstorms to strong thunderstorms and back again.

A satellite image of thunderstorms—shown as red, yellow, and orange masses—passing over the southern United States.

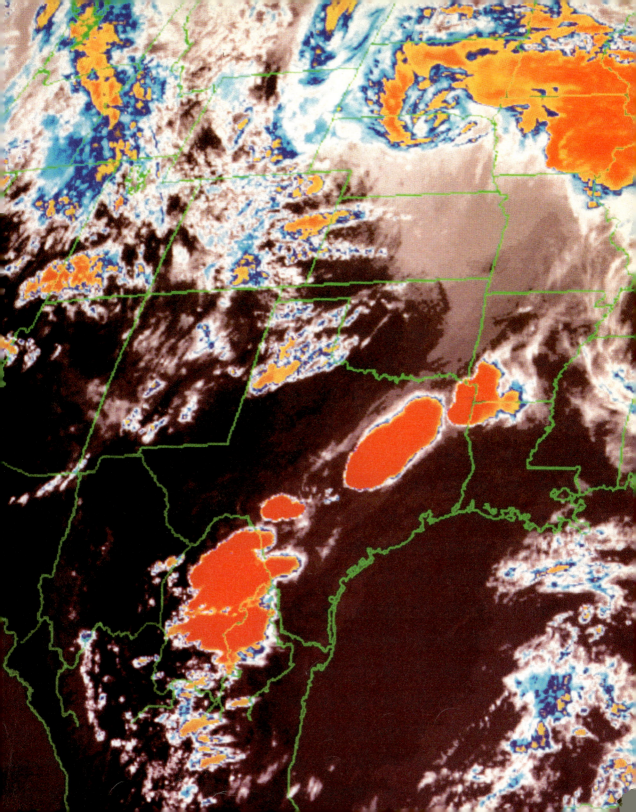

Supercell Thunderstorms

A particularly violent type of thunderstorm is the supercell thunderstorm. Like the multicellular thunderstorm, the supercell can exist for many hours. However, the supercell is not a group of many different thunderstorm cells, but one huge cell. They are often one hundred miles or more in diameter. In a supercell, changing wind directions at different heights cause the updraft into the storm to rotate. This rotation of the updraft prevents the downdraft from cutting off the storm's supply of warm, moist air. Therefore, the cell builds and grows and continues to feed off the warm, moist air for a much longer period of time than either single-cell or multicellular thunderstorms. Supercells are very powerful thunderstorms and are the source of most violent tornadoes. During multicellular or supercell thunderstorms, you must be careful if you are outdoors to protect against injury or even death.

A dangerous-looking cloud sweeps across the plains.

3 Lightning

Lightning is one of nature's most dramatic displays. In ancient times lightning and thunder were thought to represent the anger of the gods. Even today, lightning still has the power to frighten and amaze us. It is often used in horror movies to introduce especially scary scenes.

We have good reason to be afraid of lightning. Every year about one hundred people in the United States and Canada are killed by lightning. Lightning is also the most common cause of forest fires, starting more than nine-thousand each year.

Scientists cannot fully explain how thunderclouds produce lightning. However, they do know a good deal about the electrical charges in thunderclouds. Thunderclouds have areas of positive and negative electrical charges. It is thought that these charges are caused by friction between ice particles colliding with each other in the cloud. There are different sizes of ice particles in the cloud. When

Streak lightning flashes down from a night sky.

small and large particles hit each other they become electrically charged. The bigger particles, called graupel, become negatively charged. The smaller particles become positively charged. Updrafts carry the smaller, lighter particles to the upper parts of the cloud. The larger, heavier graupel stays near the bottom of the cloud.

The negative charges near the bottom of the cloud are attracted by an opposite charge on Earth's surface. When the charges become strong enough, they jump the gap between the cloud and the ground. Since electricity will flow most easily across a small gap, tall objects on the ground are better targets for lightning.

Negatively and positively charged particles formed in clouds separate and store the energy that creates lightning bolts.

Favorite Lightning Targets

Tall objects are favorite targets for lightning. Trees are especially at risk of being struck. This is because trees contain large amounts of moisture, which is a good conductor of electricity. A live tree can conduct electricity hundreds of times better than a dead tree that contains no moisture. Oak trees seem to attract lightning more than other types of trees. This is because oak trees are known to have a higher moisture content than most other trees.

When lightning hits a tree, several things can happen—the tree may start to burn, the trunk may explode, the bark may be stripped loose, or the charge may travel to the ground with little or no effect. The degree and type of damage depends on several factors. These factors include the strength of the lightning stroke, the type of lightning stroke, and the type of tree that is struck. The water in the tree, heated by an intense surge of electricity through its trunk, expands rapidly. It then turns to steam and shatters the trunk in an explosion.

The high-moisture content of this tree attracted the lightning that split it.

Hot and Cold Lightning

Strange as it may seem, lightning can be either "hot" or "cold." (Since all lightning strokes are very hot, the term "cold" is relative.) The temperature of lightning has been measured as high as 45,000 degrees Fahrenheit. This temperature is much hotter than the surface of the Sun!

While hot lightning tends to start fires, cold lightning is more likely to cause explosions. This is because hot lightning has a low flow of electrical current (the flow of electrical energy you see as a lightning bolt) that lasts for a long time. Cold lightning, on the other hand, has a high flow of electrical current that lasts for a short time. Usually, hot lightning lasts twenty- to one hundred-times longer than cold lightning. Since the length of a typical lightning flash is measured in hundredths of a second, both hot and cold lightning bolts strike quicker than you can blink. A cold bolt releases its energy so quickly, though, that an explosion occurs before a fire can even start.

Fulgurites

Have you ever heard of petrified lightning? You may have come across a piece of petrified lightning without knowing what it was. Petrified lightning looks like a hollow tube of solid sand. These tubes are called fulgurites.

Fulgurites form when lightning strikes places that have sandy soil, such as beaches and deserts. As the lightning bolt surges into the ground, it melts the sand particles in its path. This causes the sand to fuse together and form glass. Fulgurites have a rough texture and are shaped like tree branches. Most fulgurites are small, about three-quarters of an inch to three inches in diameter. Some, however, are very large. One of the biggest fulgurites ever found had two branches that were both over 15 feet long! Fulgurites can be found all over the United States and Canada. If you want to look for one, be sure to look after thunderstorms have passed through your area. Fulgurites are very fragile, so be extra careful not to break them when digging them up.

Lightning and Thunder

Lightning strikes happen very quickly. A bolt of lightning moves at the amazing speed of 31,000 miles per second. And we know that lightning is very hot. As a lightning bolt moves through the atmosphere, it heats the air it moves through. Within a fraction of a second, the temperature of the air rises by thousands of degrees. This sudden heating causes the air to expand violently. The violent expansion produces the intense sound waves that we hear as thunder. Lightning is usually seen before thunder is heard. This is because the flash of lightning and the sound of thunder move at different speeds. Light travels very fast—186,000 miles per second. Sound travels slower—1,000 feet per second. Therefore, when a lighting flash occurs, the light is seen instantly. The thunder is produced at the same time as the flash, but it takes time for the sound to reach us. Only when lighting strikes very close by do we hear the thunder at the same time.

St. Elmo's Fire

Sometimes, when thunderstorms are overhead, a bluish glow may be seen floating around objects such as flagpoles and church steeples. At night this strange glow produces eerie effects as it flickers about. The glow is called St. Elmo's fire, and it is caused by electrical discharges from the objects around which it floats.

If you ever see St. Elmo's fire nearby, you will probably hear a crackling noise, too. This noise is caused by the discharge of electricity. The appearance of

**St. Elmo's fire creates a creepy glow
around the end of this ship's yardarm.**

St. Elmo's fire means that lightning may soon strike nearby—so be careful!

4 Thunderstorm and Lightning Safety

Severe weather can be hazardous to people and property. When you are caught outdoors in a thunderstorm, you need to be particularly cautious. And even if you are indoors, such severe weather can put you at risk if high winds carry objects through the air that hit your home. A shattered window can cause cuts and bleeding that may send you to the hospital. But there are safety rules and tips that can help you avoid injury when you follow them properly.

Having a Thunderstorm Safety Plan

Thunderstorms can become severe and have the potential to cause great damage. High winds can blow down trees and damage buildings. These winds can also pick up objects such as fallen tree branches and knocked-down street signs, which are very dangerous when they're flying around

Winds from severe thunderstorms can bring down telephone and power lines.

in high winds. Lightning often accompanies thunderstorms and can be a greater danger than the winds and rain. Your family should have some plan of action or rules to follow during a severe thunderstorm.

You may want to have an NOAA (National Oceanic and Atmospheric Administration) weather radio in your home, which sounds a tone when severe weather is approaching your area. At this warning sound, you will have time to take action to help you stay safe. If you don't have one of these special radios, listen for weather bulletins and warnings on a regular radio. A battery-operated radio is best because the electricity in your home may be disabled in a serious storm.

If a thunderstorm is approaching while you are at home you should go to the basement or the lowest point in the house. Stay in the center of the house and away from any windows. If you have no basement, then the smallest room in the house with no or few windows is usually the safest. This room will probably be a closet or a bathroom. Having solid walls between you and the thunderstorm is most important. Don't stay on a second level, as high winds and lightning will increase your chances of injury.

**During a severe storm, it is best to stay indoors—
and away from the windows.**

If you are outside and see or hear an approaching thunderstorm, the safest thing you can do is get inside. If you are outside but away from your home, find a sturdy building you can go into until the severe weather passes. If there is no building, take shelter in a car.

All schools in places where thunderstorms are common should have a plan to protect students in case of severe weather. Schools also need to be equipped with an NOAA weather radio. You have probably acted in drills at school for just such an event. The inner hallways of schools are ideal for shelter from storms because of the strong brick walls between you and the storm.

Lightning Safety

There are ways to avoid, or at least decrease your risk of, being struck by lightning. Finding shelter is the most important thing you can do to avoid being struck and injured by lightning. If you are outside and cannot find immediate shelter in a building, find a cave, a valley, or an overhanging rock. Even the area beneath a bridge is a good place in an emergency. Sometimes none of these shelters are available, so you must take some other action. Find the lowest point in the area, like a ravine, and crouch down. On the sides of roads you will usually find a ditch where you can take emergency shelter.

What you don't want to do is hide beneath a lone tree in a field. Trees are tall, so they attract lightning. You should stay clear of other tall objects, such as telephone poles, even if there is a shed or overhanging object right next to it where you can take shelter. You should also stay away from metal objects, such as wire fences, metal flag poles, golf clubs, and bicycles. Metal attracts lightning easily and if it is struck, it can cause severe injury to you and others around you.

Finally, if you should happen to be in an open area and feel the hair on your head stand up, this means lightning is about to strike. Quickly crouch down and put your feet together. This makes you lower than many objects around you, and may just save your life.

Glossary

air-mass thunderstorm A thunderstorm that occurs when rising air is much warmer than the surrounding air.

atmosphere The thin layer of air that surrounds Earth.

cloudburst A short period of heavy rainfall that occurs when a cloud cannot hold its water droplets any longer and lets all of them drop at once.

condensation The process where water vapor (water in a gas form) turns into a liquid.

cumulonimbus clouds Large, gray clouds that form vertically and usually produce storms. They are also called thunderheads.

cumulus clouds Puffy white clouds associated with fair weather.

dew point The point at which the air temperature allows condensation to begin.

dissipate To break up and scatter.

downdraft A quick flow of downward air from the atmosphere.

evaporate When water turns from a liquid to a gas.

frontal thunderstorms Thunderstorms that develop along a boundary line, called a front, between a mass of cold air and a mass of warm, humid air.

fulgurites Petrified lightning that is tube-shaped and formed from melted sand.

graupel Large particles within a thundercloud that become negatively charged.

horizontal winds Winds that blow across rather than up or down.

multicellular thunderstorm A thunderstorm that has many cells that are at different stages of development.

precipitation   Any kind of water that falls to Earth from clouds, such as rain, snow, and hail.

single-cell thunderstorm   A small, weak thunderstorm that has only a single cloud.

stratosphere   A layer of the atmosphere that has slightly warmer temperatures and no clouds.

supercell thunderstorm   An especially powerful thunderstorm that is a single mass.

supercooled   Water droplets that are colder than freezing temperature (32° F) that are still in liquid form.

unstable air   Rising air that is greatly different in temperature than the air around it.

upcurrents   Winds blowing upward from beneath clouds that hold cloud droplets in the air.

updrafts   Strong vertical winds that flow up from the ground into clouds.

water vapor   Water in the form of a gas.

For Further Reading

Burby, Liza N. *Electrical Storms* (Extreme Weather). New York: The Rosen Publishing Group, 1999.

Kahl, Jonathan. *Thunderbolt: Learning about Lightning.* Minneapolis: The Lerner Publishing Group, 1993.

Sipiera, Paul and Diane Sipiera. *Thunderstorms.* Danbury, CT: Children's Press, 1998.

Resources

Nick Walker: "The Weather Dude"
P.O. Box 9535
Seattle, WA 98109
Web site: http://www.nwlink.com/~wxdude/
e-mail: wxdude@nwlink.com
KSTW-TV weather forecaster helps kids learn about weather and meteorology

Weather Education and Resources
National Weather Service Office, Portland, Oregon
5241 NE 122nd Avenue
Portland, OR 97230
(503) 261-9246
Web site:http://www.wrh.noaa.gov/portland/educate.html

WEB SITES

Dan's Wild Weather Page
by Chief Meteorologist Dan Satterfield, Newschannel 19,
    WHNT-TV Huntsville, Alabama
Web site: http://www.whnt19.com/kidwx
e-mail: webmaster@whnt19.com
Interactive weather page for kids

Enlightening Facts About Lightning
Web site: http://www.aws.com/lghtning.html

Thunderstorms and Lightning...the underrated killers
Web site:
http://www.nassauredcross.org/sumstorm/thunder1.htm
Includes environmental clues to look and listen for, and
safety tips.

Weather for Teachers and Kids
Web site:
http://info.abrfc.noaa.gov/wfodocs/weather_kids.html
Helpful weather-related links for both teachers and
students.

Index

Credits

About the Author

Dean Galiano is a freelance writer. He lives in New York City.

Photo Credits

Cover, Title Page © Angelo Hornack/Jeffry W. Myers/Corbis; pp. 6-7 © Warren Faidley/International Stock; p.8 © Cliff Riedinger/Midwestock;p.9 © Don Wolf/Midwestock; p.16 © C. Ursillo/H. Armstrong Roberts; p.18 © Michael Paras/International Stock; p.19 © Dario Perla/International Stock; pp.22-23 © David Paterson/Corbis; p.25 © Warren Faidley/International Stock; p.26 © Joann Frederick/Midwestock; pp.28-29 © Rick Adair/Midwestock; p.31 © R. Tesa/International Stock; p.35 © CORBIS/Corbis Bettman; p.36-37 © J. Anderson/H. Armstrong Roberts; p.39 © Ron Chapple/FPG; pp.10,13,15,17,30 Illustrations by Lisa Quattlebaum.

Cover Design

Kim M. Sonsky

Book Design and Layout

Lisa Quattlebaum

Consulting Editors

Mark Beyer and Jennifer Ceaser